Energy Audits

Dr. Hemant Pathak

DEDICATION

Dedicated to Shri Sainath Maharaj the all omnipotent of world the most merciful.

CONTENTS

Foreword

Glossary

Foreword

The world is moving towards a sustainable energy future with an emphasis on energy efficiency and use of renewable energy sources. A finite planet cannot support infinitely increasing consumption of resources and hence the motto of present times must be to 3 R principal - "Reduce, Reuse, Recycle".

Energy Audits; provides a unique insight into the problems our planet faces in terms of clean energy resources, and what to do about it. This books Written for academics, researchers and practitioners working in Energy field, expressed comprehensive and interdisciplinary focus on the current energy demand to enhance energy conservation outcomes.

This book has been provided to be utilized by all people concerned with energy conservation in all the industries in world.

This book provides an essential guide to researchers, it offers: various steps of Energy audits; on the challenges and experiences in present scenario.

Simply explained, Energy Audits, is an unique book bringing together diverse viewpoints from Industries and state agencies and regulators, for all who wish to make a difference in how to plan and manage our Energy resources.

Dr. Hemant Pathak
M.Sc. (Gold medalist), Ph. D.
Assistant Professor of Engineering Chemistry
Indira Gandhi Govt. Engineering College,
Sagar, MP, India

Acronymns

CH4 methane

CNG compressed natural gas

CO2 carbon dioxide

CO2e CO2 equivalent

CPP Critical Peak Pricing

CRIS Climate Registry Information System

ECMP Energy Conservation and Management Plan

EMAP Energy Management Action Plan

EMG Energy Management Group

EMS Environmental Management System

GHG greenhouse gas

GIS Geographic Information System

N2O nitrous oxide

NPV net-present value

O&M operations and maintenance

RECs renewable energy credits

REP Renewable Energy Program (Policy)

Energy units and conversion factors

Temperature Kelvin (K)

Commonly used temperature units

Celsius (C), Fahrenheit (F)

$0°C = 273.15 \text{ K} = 32°F \quad 1°F = 5/9°C \quad 1°C = 1 \text{ K}$

Fahrenheit temperature = 1.8 (Celsius temperature) + 32

Derived SI units

Heat: Quantity of heat, work, energy joule (J)

Heat flow rate, power watt (W)

Heat flow rate watt/m2

Thermal conductivity W/mK

Glossary

Abatement   The reduction or elimination of pollution.

Acid rain   The precipitation of dilute solutions of strong mineral acids, formed by the mixing in the atmosphere of various industrial pollutants

Act   A law

Aerosol   Particles of solid or liquid matter than can remain suspended in air from a few minutes to many months depending on the particle size and weight.

Air pollution   Toxic or radioactive gases   or particulate matter introduced into the atmosphere, usually as a result of human activity.

Air Toxic   Any air pollutant for which a ambient air quality standard does not exist that may reasonably be anticipated to cause cancer, developmental effects, reproductive dysfunctions, neurological disorders, heritable gene mutations or other serious or irreversible chronic or acute health effects in humans.

Ash   Incombustible residue left over after incineration or other thermal processes.

Atmosphere   The 500 km thick layer of air surrounding the earth which supports the existence of all flora and fauna.

Btu   The abbreviation for British Thermal Unit(s).

Byproduct   A secondary or additional product resulting from the feedstock use of energy or the processing of nonenergy materials. For example, the more common byproducts of coke ovens are coal gas, tar, and a mixture of benzene, toluene, and xylenes (BTX).

Carbon dioxide (CO_2)   A colorless, odorless, non-poisonous gas that is a normal part of Earth's atmosphere. Carbon dioxide is a product of fossil-fuel combustion as well as other processes. It is considered a greenhouse gas as it traps heat (infrared energy) radiated by the Earth into the atmosphere and thereby contributes to the potential for global warming. The global warming potential (GWP) of other greenhouse gases is measured in relation to that of carbon dioxide, which by international scientific convention is assigned a value of one .

Carbon sink   A reservoir that absorbs or takes up released carbon from another part of the carbon cycle. The four sinks, which are regions of the Earth within which carbon behaves in a systematic manner, are the atmosphere, terrestrial biosphere (usually including freshwater systems), oceans, and sediments (including fossil fuels)

Climate change   A regional change in temperature and weather patterns. Current science indicates a discernible link between climate change over the last century and human activity, specifically the burning of fossil fuels.

Coal
A readily combustible black or brownish-black rock whose composition, including inherent moisture, consists of more than 50 percent by weight and more than 70 percent by volume of carbonaceous material. It is formed from plant remains that have been compacted, hardened, chemically altered, and metamorphosed by heat and pressure over geologic time.

Combustion
Burning. Many important pollutants, such as sulfur dioxide, nitrogen oxides, and particulates (PM-10) are combustion products, often products of the burning of fuels such as coal, oil, gas, and wood.

Contamination
The act of polluting or making impure; any indication of chemical, sediment, or biological impurities.

Dust
Solid particulate matter that can become airborne.

Ecosystem
An interactive system that includes the organisms of a natural community association together with their abiotic physical, chemical, and geochemical environment.

Electric current
The flow of electric charge. The preferred unit of measure is the ampere.

Electric energy
The ability of an electric current to produce work, heat, light, or other forms of energy. It is measured in kilowatthours.

Electric generation
The electric generation industry includes the

industry	"electric power sector" (utility generators and independent power producers) and industrial and commercial power generators, including combined-heat-and-power producers, but excludes units at single-family dwellings.
Emission	Release of pollutants into the air from a source. We say sources emit pollutants. Continuous emission monitoring systems (CEMS) are machines, which some large sources are required to install, to make continuous measurements of pollutant release.
Emissions coefficient	A unique value for scaling emissions to activity data in terms of a standard rate of emissions per unit of activity
Electricity	A form of energy characterized by the presence and motion of elementary charged particles generated by friction, induction, or chemical change.
Energy	The capacity for doing work as measured by the capability of doing work (potential energy) or the conversion of this capability to motion (kinetic energy).
Energy consumption	The use of energy as a source of heat or power or as a raw material input to a manufacturing process.
Energy Audit	An assessment of a home's energy use. These include a number of different types of surveys,

including (in increasing order of cost and complexity): online audits, in-home home energy surveys, diagnostic home energy surveys, and comprehensive home energy audits.

Energy Conservation

Saving energy by doing with less or doing without (e.g., setting thermostats lower in winter and higher in summer; turning off lights; taking shorter showers; turning off air conditioners; etc.).

Energy Efficiency

A ratio of service provided to energy input. Services provided can include buildings-sector end uses such as lighting, refrigeration, and heating: industrial processes; or vehicle transportation. Unlike conservation, which involves some reduction of service, energy efficiency provides energy reductions without sacrifice of service.

Energy loss

Deleted because there is no need for a general term to encompass all forms of energy loss. Terms referring to losses specific to particular energy sources are defined separately.

Exposure

The concentration of the pollutant in the air multiplied by the population exposed to that concentration over a specified time period.

Fossil fuels

Fuels such as coal, oil, and natural gas; so-called because they are the remains of ancient plant and animal life.

Global warming   increase in the average temperature of the earth's surface.

Greenhouse gases   Atmospheric gases such as carbon dioxide, methane, chlorofluorocarbons, nitrous oxide, ozone, and water vapor that slow the passage of re-radiated heat through the Earth's atmosphere.

Hydrocarbons   Compounds containing various combinations of hydrogen and carbon atoms. They may be emitted into the air by natural sources (e.g., trees) and as a result of fossil and vegetative fuel combustion, fuel volatilization, and solvent use. Hydrocarbons are a major contributor to smog.

Industrialized countries   Nations whose economies are based on industrial production and the conversion of raw materials into products and services, mainly with the use of machinery and artificial energy (fossil fuels and nuclear fission).

Kyoto Protocol   An international agreement adopted in December 1997 in Kyoto, Japan. The Protocol sets binding emission targets for developed countries that would reduce the emissions on average 5.2 percent below 1990 levels.

Micro- (μ)   The metric prefix for one millionth of the unit that follows.

Microgram (μg)   One millionth of a gram: 1 μg = 10^{-6} g = 0.001 mg.

Mitigation   Actions taken to avoid, reduce, or compensate for the effects of environmental damage. Among the broad spectrum of possible actions are those that restore, enhance, create, or replace damaged ecosystems.

Monitoring   Periodic or continuous surveillance or testing to determine the level of compliance with statutory requirements and/or pollutant levels in various media or in humans, plants, and animals.

Non-point pollution   diffuse pollution mainly from agriculture or dumping grounds. It is difficult to collect for treatment.

Opacity   The amount of light obscured by particle pollution in the atmosphere. Opacity is used as an indicator of changes in performance of particulate control systems.

ppb/ ppm   Units commonly used to express contamination ratios, as in establishing the maximum permissible amount of contaminant in water, land, or air.

Plume   A visible or measurable discharge of a contaminant from a given point of origin that can be measured according to the Ringelmann scale.

Pollutants (pollution)   Unwanted chemicals or other materials found in the air. Pollutants can harm health, the environment and property. Many air pollutants occur as gases or vapors, but some are very tiny solid particles: dust, smoke, or soot.

Point pollution   polluted water from a defined point. It can be collected as industrial or municipal wastewater and treated by what is often called end-of-pipe technology (environmental technology).

Pollution control
The addition of processes, practices, materials, products or energy to waste streams to reduce the risk posed by pollutants and waste before their release to the environment.

Pollution prevention
The use of processes, practices, materials, products, substances or energy that avoid or minimize the creation of pollutants and waste, and reduce the overall risk to human health or the environment

Public health
the health or physical well-being of a whole community.

Reuse
The reemployment of products or materials, in their original form or in new applications, with refurbishing to original or new specifications as required.

Risk assessment
Methods used to quantify risks to human health and the environment.

Smog
A mixture of pollutants, principally ground-level ozone, produced by chemical reactions in the air involving smog-forming chemicals.

Solid waste
non-liquid, non gaseous category of waste from non-toxic household and commercial sources.

Toxic emissions
poisonous chemicals discharged to air, water, or land.

Toxic waste
garbage or waste that can injure, poison, or harm living things, and is sometimes life-threatening.

Visibility

A measurement of the ability to see and identify objects at different distances. Visibility reduction from air pollution is often due to the presence of sulfur and nitrogen oxides, as well as particulate matter.

Waste

Garbage, trash.

1. Introduction

Energy is the driver of growth. Energy is one of the most important building block in human development, and essential factor in determining the economic development of any nation. Energy, irrespective of its form is a scarce commodity and a most valuable resource.

Energy Audit is the key aspect of energy conservation and management. It is the systematic approach for decision making in the areas of energy management.

International studies on human development indicate that India needs much larger per capita energy consumption to provide better living conditions to its citizens.

India's second largest population and increasing pace of economic growth make its energy needs particularly challenging.

Resource augmentation and growth in energy supply have failed to meet the ever increasing demands exerted by the multiplying population, rapid urbanization and progressing economy.

To determine how and where energy is being used or converted from one form to another, to identify opportunities to reduce energy usage, focus on the techniques used are intended to get the picture of energy balance in a facility – the inputs, uses and losses, for evaluate the economics and technical practicability of implementing these reductions and to formulate prioritized recommendations for implementing measures, energy audit is an mandatory step.

Energy shortages continue to plague India, forcing it to rely heavily on imports consumption of coal. Electricity is the commonly used form of energy. If the use of this is managed correctly it can therefore help economy cut their energy consumption and costs.

Energy is one of the most important resources to sustain our lives. At present we still depend a lot on fossil fuels and other kinds of non-renewable energy.

The extensive use of renewable energy including solar energy needs more time for technology development.

Energy Conservation is the critical needs in any countries in the world. But such growth has to be balanced and sustainable.

While Energy audit may be defined as "The Verification, Monitoring and Analysis of use of energy including submission of Technical Reports containing recommendations for improving energy efficiency with cost benefit analysis and an action plan to reduce energy consumption".

India's energy requirement comes from five sectors; agriculture, industry, transport, services and domestic, each having considerable saving potential. For example, energy costs amount to 20 percent of the total production cost of steel in India which is much higher than the international standards.

The energy intensity per unit of food grain production in India is 3 – 4 times higher than that in Japan. Sustainable growth also implies that our energy management and energy conservation measures are eco-friendly and accompanied by minimum pollution, in particular minimum carbon emission.

Human population and the individual life expectation will increase, energy could, in the future, be in short supply. Unless that supply is increased, it will be a source of friction in human affairs Energy Conservation is the deliberate practice or an attempt to save electricity, fuel oil or gas or any other combustible material, to be able to put to additional use for additional productivity without spending any additional resources or money.

Energy Audit is an effective tool in defining and pursuing comprehensive energy management. It attempts to balance the total energy inputs with its use and serves to identify all the energy streams in a facility.

It quantities the energy usage according to its discrete functions. It helps to calculate the energy cost reduction, preventive maintenance and quality control programme which are vital for production and utility activities.

Energy audits consist of the basic principal of management like planning, decision making, organizing and controlling, apply equally as in any other management subject.

2. Energy Audit

Energy Audit estimated the ways energy and fuel are used in any nations, and help in identifying the areas where waste can occur and where scope for improvement exists. Also record of energy used for comparison against a budget or another standard of performance.

Conducting regular energy audits, where auditors inspect, analyze and evaluate energy consumption, allows to assess how much energy their uses and to pinpoint opportunities for potential energy and cost savings.

An audit is a vital planning and communicating tool that ensures consistency and completeness of audit coverage of the subject matter and effective use of resources.

however, if implement their auditors' recommendations.

The audit plan should spell out the following:

• Details of the auditee

• Dates and places

• Audit objectives

• Scope and criteria

• Identification of the energy audit elements

• Expected time and duration

• Audit team members

• Schedule of meetings

• Confidentiality requirements; and

• Audit report content with date of issue and distribution.

The preparation of the audit plan is the duty of the principal auditor. The plan should be communicated to all parties concerned, i.e. the client the audit team and the auditee.

3. Scope

Energy Audit is the appropriate reflection of conservation strategy into realities, by mixing technically conclusion with economic and other organizational requirements within national laws and regulations and specified time frame.

The audit scope describes the extent and boundaries of the audit in terms of factors such as physical location and organizational activities, as well as manner of reporting.

Also determines the current situation and the base on which energy efficiency improvements will be built and to get a detailed analysis of energy use and losses in a specific process or facility. The client must sit down with the auditor and establish the scope of the work. Set audit boundaries limiting the scope of an audit in a large, complex facility.

It may help to visualize the audit boundary enclosing the audit area and then to focus on the energy streams flowing into and out of the box.

Measurement of these energy flows may play a large part in determining the audit boundary.

4. Steps of Energy Audit

The general energy audit consists of four main stages:

4.1 Initiating

Any Nation or establishment takes this first step early in the process of establishing an energy management procedure. As such, those ordering an audit are in the role of a client and will normally be the recipients of the final audit report. The auditee is any establishment to be audited.

4.2 Preparing

Preparation of energy audit's include the auditee's staff size, the staff 's capability and availability, the outside consultant's capability, and money and time available.

Attempts to stretch the audit's scope beyond any of these resources may compromise the quality of the audit. Audit quality should never be sacrificed in pursuit of greater or new area.

For an successful energy audit it should be approached in the spirit of collaboration. Members of the auditee's staff need to feel that they are participating positively in the process. The auditor should consult the auditee about the scope of the audit, seek information regarding areas of concern that need priority consideration and discuss the planned audit methodology, among other tasks.

A familiarization visit to the facility before proceeding with other audit preparations serves several purposes: personal contacts and lines of communication are established; a clearer picture of the facility and the scope emerges; issues may be clarified; resources may be identified and secured; and adjustments to the planned audit scope, date and duration may be made.

The auditee is a valuable source of criticism for the audit program. The insights gained can greatly improve the audit process and help to produce better-quality results.

Also, pre-audit checklists may be administered during the pre-audit visit. This will help to minimize the time spent at the site during the actual energy audit and maximize the auditor's productivity. On-site time is costly for both the auditor and the operation being audited.

An audit work is a important outcomes that must be flexible enough to permit immediate adjustments to emphasis on account of information gathered during the audit or changed conditions.

Using an audit checklist or the term audit protocol may ensures that the goal will be reached in the minimum amount of time and that no important points of the journey will be missed.

4.3 Executing

To assess the overall plant and to establish the scope of the remaining work, including investigation and analysis. The following are the key tasks mandatory:

• Reconcile utility data with operating information;

• Identify the areas of energy consumption;

• Establish a work plan for gathering information at the audit site, analyzing all data and producing an audit report.

During the audit through interviews and the examination of records and observations the checklists are used to identify problem areas. These are to be examined more closely in elaborated detailed, diagnostic audits.

* How Much Energy Used and Consumed?

* How is It Used and Can We Reduce Cost/ Consumption?

* How to calculate Losses / Reduce Losses?

* Benchmarking For Various Processes/Systems?

1. Analyse present consumption and past trends in detail

2. Review energy uses requirements

3. Consider sub-metering

4. Compare standard consumption to actual

5. Produce an energy balance diagram for the establishment

6. Review existing energy recording systems

7.Compare consumption with other locations, other establishments, previous period, norms.

8. Check capacities and efficiencies of equipment.

9. Users training and review projects.

10. Management information system with energy parameters.

11. Develop energy use indices to compare performance/ productivity.

12. Monitoring procedures and examine new energy saving techniques.

13. Examine need for energy saving incentives and publicity campaign strategies.

The opening meeting sets the tone of the audit. Spend time and effort on the opening meeting, they will have confidence in the results. Audit team members and the facility staff meet, perhaps for the first time, so as to

- Describe audit methodologies and define communications links

- Review the purposes, scope and plan of the audit; make changes to the audit plan as required;

- Availability of resources and facilities; confirm the schedule of meetings

• To inform the audit team about relevant site health and safety and emergency procedures; answer questions;

• establish a comfort level between the two groups.

The auditee's staff is encouraged to participate actively in the audit and keep notes on their own observations.

The principal auditor should also point out the limitations of the audit, the chief one being that the examination is based on limited-time observations.

On the basis of available information, whether the energy survey of the facility will be carried out by examining either the entire facility area by area; or the various energy-using systems one at a time.

4.4   Reporting results

An Audit report explained decision of conducting energy audit followed audit scope, objective and criteria. Report contained a fruitful advise to the auditee   after preliminary analysis from Secure resources. After detailed analysis and evaluation this reports consist of various checklist and working documents.

This draft formulate recommendations reviewed by auditee's representative. The data collected during the general and diagnostic audits are used to calculate the amounts of energy used in, and lost from, equipment and systems.

By calculating the value of this energy, the auditors produce more accurate estimates of the savings to be expected from an energy project.

Analysis of the energy surveys will indicate the energy services with the most potential for immediate improvement. A cost-benefit analysis based on future energy costs will show the merit of each potential improvement and help to set priorities.

It may contained-

✓ Accounts for energy input, storage, conversion, waste, sales and consumption.

✓ Measures instruments, calibration and measuring rate. This includes the scope of energy report, frequency of submission, breakdown level, analysis, etc.

✓ Assesses to what extent company equipment makes efficient Prepare audit plan

After launching of an audit report conduct opening meeting and conduct initial walk-through tour for checking audit facility as per plan with Collect information.

May be carry out diagnostic audit(s) with analyse information to Evaluate audit findings. Review EMOs with auditee's representative and conduct closing meeting.

Toward the close of the audit, all information gathered during the audit is reviewed by the auditors, and tentative findings and observations are formulated.

Audit report can be finalized and distributed to the client and the auditee. The auditor's conclusions should not be influenced by considerations of such factors as impact on business units or schedules of production.

Only way to ensure the independent, unbiased and fresh view of the auditee's operations is to use an independent consultant or staff from other business units.

Audit checklist

Establishment: _____

Address: _____

Audit location: _____

Audit objective(s): _____

Scope – boundaries: _____

Audit criteria: _____

Areas to be examined

- Entire site
- Buildings
- On-site services
- Lighting and Heating
- Individual services
- Boiler plant
- Distribution systems
- Domestic and process water
- Process refrigeration
- Production and process operations
- Electrical and Other
- Ventilating and air conditioning
- Rate structures

Resources

- Staff
- Technical and Clerical
- External
- Consultants
- Utility companies
- Service companies
- Government organizations
- Contractors

Measuring and monitoring equipment available

Describe: _____

Building characteristics

Remaining life of Building structure: _____ years

Envelope system: _____ years

VAC system: _____ years

Interior partitions: _____ years

Changes and renovations planned (details): _____

Building conditions

- Current problems
- Appearance and Comfort
- Breakdowns and Lack of capacity
- Noise and Other parameters

Investment and operational needs and desires

☐ Save energy

☐ Reduce use of fuel: _____

☐ Reduce time systems operating under maximum-demand conditions

☐ Accommodate increased load

☐ Charge energy costs directly to consumers

☐ Reduce requirement for manual operation

Will audit recommendations be applied to other buildings?

- Yes
- No

Explain:_____

Deadlines

Audit dates (from, to): _____

Date completion required: _____

Date preliminary findings required: _____

Date final report required: _____

Report distribution: _____

Implementation

Housekeeping deadlines: _____

Low-cost deadlines: _____

Financial limits: _____

Retrofit deadlines: _____

Financial limits: _____

Reporting format

Level of detail required: _____

Financial analysis required: _____

Acceptable payback period: _____

Tax advantages

Details: _____

Grants and subsidies available

Details: _____

Agreements

Organization's representative, name, title, signature:_____

Date: _____

Principal auditor – name, company, signature:_____

Date: _____

5. References

1. Energy for a sustainable world: Jose Goldenberg, Thomas Johansson, A. K. N. Reddy, Robert Williams (Wiley Eastern).

2. Modeling approach to long term demand and energy implication : J.K.Parikh.

3. Energy Policy and Planning : B.Bukhootsow.

4. World Energy Resources : Charles E. Brown, Springer2002.

5. 'International Energy Outlook' -EIA annual Publication

6. Principles of Energy Conversion: A.W. Culp (McGraw Hill International edition.)

7. Aspects of Energy Conversion : I.M.Blair and B.O.Jones

8. Principles of Energy Conversion : A.W.Culp (McGrawHill International)

9. Energy conversion principles : Begamudre , Rakoshdas

10. Principles of Energy Conversion : A.W. Culp.

11. Energy Management: W.R.Murphy, G.Mckay (Butterworths).

12. Energy Management Principles: C.B.Smith (Pergamon Press).

13. Efficient Use of Energy : I.G.C.Dryden (Butterworth Scientific)

14. Energy Economics -A.V.Desai (Wieley Eastern)

15. Industrial Energy Conservation : D.A. Reay (Pergammon Press)

16. Energy Management Handbook – W.C. Turner (John Wiley and Sons, A Wiley Interscience Publication)

17. Industrial Energy Conservation Manuals, MIT Press, Mass, 1982

18. Energy Conservation guide book Patrick/Patrick/Fardo (Prentice Hall)

19. CRC Handbook of Energy Efficiency – CRC Press.

20. Efficient Use of Energy: I.G.C.Dryden (Butterworth Scientific)

21. Industrial Energy Conservation: D.A. Reay (Pergammon Press)

22. Energy Management Handbook – W.C. Turner (John Wiley and Sons, A Wiley Interscience publication)

23. Industrial Energy Management and Utilisation –L.C. Witte, P.S. Schmidt, D.R. Brown (Hemisphere Publication, Washington, 1988)

24. Industrial Energy Conservation Manuals, MIT Press, Mass, 1982

25. Energy Conservation guide book Patrick/Patrick/Fardo (Prentice hall1993)

ABOUT THE AUTHOR

Dr. Hemant Pathak held positions as Assistant Professor in the department of chemistry, Govt. Indira Gandhi Engineering College, Sagar, MP, India. He had extensive experience in teaching, research and administrative management.

Dr. Pathak received his Ph.D. degree in chemistry from Dr. Hari Singh Gour Central University, Sagar, India and M.Sc. Gold medalist from Jiwaji University, Gwalior. He has published 18 books and more than 50 research papers in reputed International and National journals and received several awards. He is a member of editorial boards and reviewer boards of several international journals and societies. His area of specialization includes Engineering Chemistry and Environmental Pollution management.

www.ingramcontent.com/pod-product-compliance
Lightning Source LLC
Chambersburg PA
CBHW071558170526
45166CB00004B/1709